Training with the Master

Morihei Ueshiba (1883–1969), Founder of Aikidō—the Way of Harmony and the Art of Peace.

TRAINING WITH THE MASTER

Lessons with Morihei Ueshiba, Founder of Aikidō

John Stevens *and*
Walther v. Krenner

SHAMBHALA
Boston & London
2004

Shambhala Publications, Inc.
Horticultural Hall
300 Massachusetts Avenue
Boston, Massachusetts 02115
www.shambhala.com

© 1999 by John Stevens and Walther v. Krenner

All rights reserved. No part of this book may be reproduced
in any form or by any means, electronic or mechanical, including
photocopying, recording, or by any information storage and
retrieval system, without permission in writing from the publisher.

9 8 7 6 5 4 3 2 1

First Paperback Edition
Printed in the United States of America

⊗ This edition is printed on acid-free paper that meets the
American National Standards Institute z39.48 Standard.
Distributed in the United States by Random House, Inc.,
and in Canada by Random House of Canada Ltd

The Library of Congress Catalogues the previous edition of this book
as follows:
Stevens, John, 1947–
Training with the master: lessons with Morihei Ueshiba, founder of
Aikidō/John Stevens and Walther v. Krenner.—1st ed.
p.   cm.
Includes bibliographical references (p.).
ISBN 1-57062-568-9 (paper)
ISBN 1-57062-322-8 (cloth: alk. paper)
1. Ueshiba, Morihei, 1883–1969.   2. Martial artists—Japan—Biography.
3. Martial artists—Japan—Biography—Pictorial works.   4. Aikido.
5. Aikido—Pictorial works.   I. Krenner, Walther v.   II. Title.
GV1113.U37S96   1999   98-20505
796.815'4'092—dc21   CIP

Contents

Preface
vii

Introduction
ix

About the Photographs
xxiii

PART ONE
On the Mat with the Master
1

PART TWO
The Master Teaches the Way
79

PART THREE
The Art of Peace
113

Suggestions for Further Reading
139

Preface

by John Stevens

TRAINING WITH THE MASTER is a photo-essay depicting the training methods and daily life of Morihei Ueshiba, perhaps the greatest martial artist who ever lived, at age eighty-four, the summit of his career as a spiritual seeker. After a long and arduous quest for the true meaning of enlightened warriorship, Morihei was inspired to create Aikidō, the Way of Harmony and the Art of Peace, a revolutionary art of physical and spiritual integration. Morihei is one of the modern world's most important spiritual teachers. His grand vision of Aikidō as a vehicle of liberation from all partial, limiting views of the world is as profound and compelling as the philosophy of Krishnamurti, and his teachings on nonviolence are as admirable and universally applicable as the concepts of *ahimsa* and *satyagraha* advocated by Gandhi.

Over the years I have examined thousands of photographs of Morihei Ueshiba, but when I viewed the set of photographs reproduced in this book, I was enthralled by the special luminous quality of each shot. There is a real magic in this particular set of photographs, a subtle bond that ties them all together even though they were shot at different times in 1967, from various angles, and in several different locales, and not one was arranged or posed. I am greatly pleased that this wonderful visual record of Morihei Ueshiba will now be made available to the world at large.

The photographs in part one, "On the Mat with the Master," show Morihei giving instruction, at various times, in the techniques and philosophy of Aikidō in the old Hombu (Headquarters) Dōjō in Tokyo. Gazing at these photographs, it is not difficult to imagine being on the mat with the master, a precious opportunity for all those who did not have the opportunity to actually encounter Morihei in the flesh. Part two, "The Master Teaches the Way," depicts Morihei in everyday life—entering a room, sitting in a relaxed yet dignified manner, chatting easily and joking with his students and other visitors, making a serious point, taking refreshment—as he embodies the Way of Harmony naturally in all his actions. The auspicious sight of such vivid images makes a very strong impact; a glimpse of a beatific smile or a view of a simple but graceful gesture can teach us much about the Art of Peace. I have included a short biography of Morihei as an introduction (the illustrations in this section are courtesy of Kisshōmaru Ueshiba), but I have kept the captions to the photographs in parts one and two as brief as possible, preferring to let each striking image speak for itself.

Part three, "The Art of Peace," contains the essence of Morihei's philosophy, distilled from his collected talks and oral tradition. The text originally appeared in *The Art of Peace,* published as one of the volumes in the Shambhala Pocket Classics series. Morihei's sayings are truly brought to life augmented with the splendid photos of parts one and two showing the master manifesting the Art of Peace on and off the mat. The book concludes with suggestions for further reading for those who wish to have more detailed information on Morihei's life and teaching.

Introduction

by John Stevens

MORIHEI UESHIBA was born in Tanabe City, Japan, an old castle town and seaport about two hundred miles south of Ōsaka, on December 14, 1883. He was the fourth child (out of five) and the only son born to Yoroku and Yuki Ueshiba. The Ueshiba clan was one of the more prominent families in the area, and a number of Morihei's male relatives were famed as strong men. Morihei, though, was born a bit prematurely and was somewhat frail as a child. He grew into a robust teenager, however, building up his strength and stamina by swimming in the bay, working on the fishing boats, hiking in the mountains, and competing in every sumo contest held in the area.

Around age six, Morihei was sent to a temple school. He did not care much for the dry Confucian classics, but he was enthralled by the esoteric rituals and secret chants the schoolmaster, a Shingon Buddhist priest, taught him. Tanabe is located in the Kumano district, site of many holy shrines and magnificent Buddhist temples, and the young Morihei was nourished in a rich mystical milieu.

Morihei loved to read and devoured hundreds of books on all subjects, but he did not like being cooped up in a classroom and he spent only a year in a regular public school as a teenager. After he dropped out of public school, Morihei enrolled in a private abacus academy. Following graduation from the academy, he took a job in the local tax office. The work did not suit Morihei—frequently he sided with the hard-pressed taxpayers against the government—and in 1901 Yoroku gave his eighteen-year-old son the address of some prosperous relatives, supplied him with some capital, and sent the young man to Tokyo to seek his fortune. Morihei did well there at first, establishing a small stationery business, but his heart was not in business, and city life made him ill; within a year he was back in Tanabe.

Although Japan had been shut off from the rest of the world for nearly two and a half centuries, once the country was forced to confront the West, it embarked on a breakneck buildup of a modern military force, initially to prevent colonization and then, emulating the Western powers, to engage in some imperialism of its own. Japan fought and won a war with China in 1894–95, and by the beginning of the twentieth century the island nation was ready to take on continental colossus Russia. In 1903, not long after marrying Hatsu Itogawa, Morihei received a summons to serve in Japan's army. Morihei failed the initial induction examination because he didn't meet the minimum height requirement of five feet two inches. Mortified by the rejection, Morihei attached heavy weights to his legs and hung from trees for hours to stretch his spine the extra half inch. Morihei passed the next physical and was assigned to a regiment stationed in Ōsaka.

Fiercely competitive and intensely driven, Morihei relished the harsh discipline of boot camp. He was relentless on forced marches, determined to finish first with the added burden of extra backpacks picked up from stragglers; he volunteered for every base construction project, single-handedly lifting the heaviest of loads and displacing huge boulders; and he quickly established himself as camp sumo champion and top bayonet fighter. Morihei had dabbled briefly in the martial arts when he was living in Tokyo, but while stationed in Ōsaka he entered the Gotō-ha Yagyū-Ryū, headed by Masakatsu Nakai, and received his first systematic instruction in classical warrior arts.

Despite his prowess as a fighting man, when war finally erupted between Japan and Russia in February 1904, Morihei was not immediately called up, and even after he was sent to the mainland he was largely kept away from the front. Morihei's father had secretly contacted the military authorities and pleaded with them not to risk the life of his only son. Morihei, however, was close enough to the fighting to witness the dreadful carnage and later remarked, "War always means death and destruction and that is never a good thing."

Upon the conclusion of the war, in Japan's favor, in 1905, Morihei's superiors wished to sponsor the gung ho soldier for officers school, but Morihei spurned that valuable recommendation and returned to civilian life in Tanabe in 1906. For the next few years Morihei seemed to be at a complete loss; he was moody and uncommunicative to his family and friends, and he disappeared for days, either by shutting himself up in his room to fast and pray or by hiding out in the mountains. During this troubled period Morihei continued his study of the martial arts, however, receiving a diploma from the Gotō-ha Yagyū-Ryū, and in 1909 he came under the beneficial influence of Kumagusu Minakata (1867–1941), an eccentric philosopher who had settled in Tanabe.

The polymath Kumagusu—who had lived in the West for sixteen years and established himself as a world-class savant of natural science, folklore, and religion—is honored today as the father of Japan's environmental movement. The perceptive Kumagusu had clearly recognized, a century ago, the pernicious effects of unbridled development and the dangers of industrial pollution. When Morihei came into Kumagusu's sphere, Kumagusu was campaigning against the Shrine Consolidation Policy promulgated by the government in 1906, a plan to force smaller shrines to consolidate with larger ones, with the government seizing the "excess" property for development. Kumagusu argued that such a policy would result in the destruction of natural wildlife sanctuaries and essential watersheds, and that it would seriously disrupt traditional village life and adversely affect the local economy. Morihei teamed up with Kumagusu to spearhead the protest movement, which proved to be quite successful—Tanabe lost only a few shrines, and the entire Kumano district remained largely untouched by either public or private development.

Kumagusu encouraged Morihei to seek out new challenges, and in 1912 Morihei and eighty-four other pioneers from Tanabe headed for the distant frontier province of Hokkaidō to settle in the district of Shirataki. Being a pioneer in the frigid north was no easy task—the crops failed during the first three years of the settlement and the group had to subsist on mountain vegetables, nuts, and river fish—but Morihei worked tirelessly to build a model community in the wilds of Hokkaidō. A disastrous fire in 1916 that destroyed nearly eighty percent of the village was a great setback

Thirty-four-year-old Morihei (seated center) with other members of the Shirataki village council in 1917. In his thirties, Morihei was a stern and serious model citizen.

and the number of settler families dwindled from fifty-five to thirty in eight years, but Morihei never wavered in his efforts to establish a viable community. In 1917, he was elected to serve on the village council.

In Hokkaidō, Morihei continued his practice of the martial arts, taking on all comers in sumo contests and wooden bayonet matches, and dealing with the many highwaymen that roamed the solitary Hokkaidō roads. To increase his physical power, Morihei wrestled huge logs he had cut down with a specially weighted ax—in one season Morihei felled and chopped up five hundred trees on his own. Morihei was supremely confident of his muscular strength and technical skill until he encountered the legendary Sōkaku Takeda, grandmaster of the Daitō-Ryū.

Although barely five feet tall and slight of build, Sōkaku (1859–1943) had acquired a fearsome reputation as a warrior. Born into the Takeda clan of

Aizu, a family renowned for its proficiency in fighting arts, Sōkaku honed his skills by traveling all over Japan, engaging in hundreds of contests, some fought with live blades, against all manner of martial artists and street fighters. When Morihei first met Sōkaku in Hokkaidō in March 1915, the fifty-six-year-old Daitō-Ryū master was conducting a seminar at an inn. Morihei was completely dominated by Sōkaku and immediately requested that Sōkaku accept him as a student. Thereafter, Morihei trained with Sōkaku as much as possible, accompanying the master on teaching tours and inviting him to stay and instruct at Morihei's residence in Shirataki.

Morihei's stay in Hokkaidō had been valuable for him—he had flourished in the vast wilderness, helping to create a thriving village from scratch, had built up tremendous strength and stamina working outdoors, and had received personal instruction from the top martial artist of the day—but he was still restless, searching for something more than material success and prowess as a martial artist. In December 1919, when word arrived that his father Yoroku was deathly ill back in Tanabe, Morihei, without hesitation, gave his house and property to Sōkaku and left Hokkaidō for good. (Morihei met Sōkaku a few times after this, but as he embarked on a new course, Morihei gradually distanced himself from Sōkaku and the Daitō-Ryū.)

Morihei did not return directly to Tanabe to be at his father's side, however. Something drew him instead to Ayabe, home of the Ōmoto-kyō, a dynamic new religion led by the grand shaman Onisaburō Deguchi (1871–1947). Morihei was immediately smitten with Onisaburō's spiritual teachings, which emphasized the innate divinity of each and every human being, and he longed to heed the call of Ōmoto-kyō to "reform the world and create heaven on earth."

By the time Morihei finally returned to Tanabe, Yoroku had already passed away—as Onisaburō had predicted: "Your father is ready to leave in peace. Let him go." After a brief period of spiritual turmoil, early in 1920 Morihei announced his decision to take his family and his mother with him to Ayabe.

After settling in a small house near the main shrine on the Ōmoto-kyō compound, Morihei assisted with the many farming and construction projects then in full swing, as well as participating in the various prayer services, meditation sessions, special fasts and feasts, and purification ceremonies. Onisaburō also requested that Morihei open a training hall to teach Budō (martial arts) to the other Ōmoto-kyō believers.

Morihei was pleased to be in Ayabe, but the first year there was very trying for the Ueshiba family. Morihei lost both his sons, Takemori (age three) and Kuniharu (age six months), to illness in 1920, and in February 1921 the Ayabe compound was raided by government agents.

For some time the government had been monitoring Ōmoto-kyō activities. Onisaburō's pacifist, anticapitalist, and anti-imperialist views, coupled with the growing popularity of Ōmoto-kyō, alarmed the government sufficiently to have Onisaburō arrested on the charge of lèse-majesté and to order the ransacking of the Ōmoto-kyō headquarters.

Morihei was not seriously affected by this first Ōmoto-kyō incident, since he was a new member and not under government surveillance. For the next two years, Morihei quietly devoted himself to his Ōmoto-kyō studies with Onisaburō (who was released on bail bond after four months' imprisonment), organic farming, and martial art training. His third and sole surviving son, Kisshōmaru, was born in June 1921. (Morihei also had a daughter, Matsuko, born in 1911 in Tanabe.)

In February of 1924, Onisaburō, his bodyguard Morihei, and two other Ōmoto-kyō believers secretly embarked on the great Mongolian adventure, a journey to locate the secret city of Shambhala on the continent. From that base, Onisaburō dreamed of setting up a heavenly kingdom on earth, a new Jerusalem, and then he would organize the peoples of East and West into a universal association of love and fellowship. After many adventures and several face-to-face encounters with death at the hands of bandits and the Chinese army, Onisaburō had to abandon his grand scheme—"The timing was not right," he remarked without regret—and the group returned to Japan. (Onisaburō was immediately placed in detention for violating the provisions of bail.)

The severe trials and tribulations of the great Mongolian adventure had a profound effect on Morihei. His life was in constant danger, and several times he had fought off, in deadly hand-to-hand combat, murderous attacks by cutthroat bandits, and their group narrowly escaped execution by the Chinese army—reprieve came just before they were to be placed in front of a firing squad. He was particularly affected by this incident:

> Once we were trapped in a valley and showered with bullets. Miraculously, I could sense the direction of the projectiles—beams of light indicated their paths of flight—and I was able to dodge the bullets. The ability to sense an attack is what the ancient masters meant by anticipation. If one's mind is steady and pure, one can instantly perceive an attack and avoid it. This sixth sense is the essence of *aiki* [techniques of harmonization].

Upon his return to Ayabe, Morihei was a changed man. He trained harder than ever in the martial arts and spent days in seclusion in the mountains engaging in spiritual exercises. In the spring of 1925, the forty-two-year-old Morihei was transformed by an intense vision. After defeating a high-ranking swordsman who had come to challenge him, Morihei went into his garden to towel the sweat from his face.

> Suddenly the earth trembled. Golden vapor welled up from the ground and engulfed me. I felt transformed into a golden image, and my body seemed as light as a feather. All at once I understood the nature of creation: the way of a warrior is to manifest divine love, a spirit that embraces and nurtures all things. Tears of gratitude and joy streamed down my cheeks. I saw the entire earth as my home, and the sun, moon, and stars as my intimate friends. All attachment to material things vanished.

Following this enlightenment experience, Morihei manifested powers that can only be described as miraculous. He could throw ten attackers at once, pin huge sumo wrestlers with

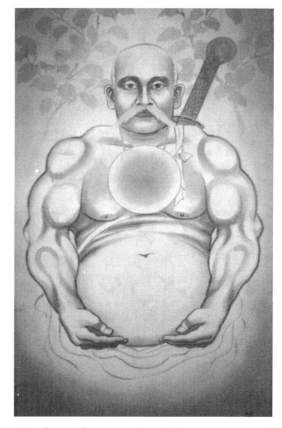

Morihei in his transfigured form as master of the Art of Peace. Following his enlightenment experience in the spring of 1925, Morihei emerged as an invincible warrior capable of performing astounding feats.

Introduction • xv

a single finger, even dodge bullets when challenged by a military firing squad. (Many of these amazing feats were recorded on film.) Not surprisingly, Morihei's incredible skill as a martial artist attracted national attention. Soon Morihei had a wide circle of influential supporters who urged him to move to Tokyo and establish a dōjō. In 1931 Morihei opened the Kōbukan in Tokyo, where he gave instruction in Aiki-Budō to Japan's elite, becoming one of the most important figures in the nation.

This was a period of great political and social turmoil in Japan, however, and even though Morihei had largely separated himself from the Ōmoto-kyō organization—with Onisaburō's blessing: "Your duty is to teach the real meaning of *budō* to the entire world"—Morihei was implicated in the second Ōmoto-kyō incident in 1935, a much harsher repression of the sect by the government than the first one in 1921, and a warrant was issued for his arrest. Morihei's supporters in the police agency were able to prevent his detention in jail, but it seems that Morihei was under house arrest for some time. (Onisaburō was to spend more than six years in prison.) Morihei remained under military police surveillance, and it would have made things easier for him to take his friends' advice and denounce Onisaburō, but to Morihei's credit he refused: "Onisaburō is my teacher and I will never renounce him to save my own skin."

Japan was gearing up for world war during this period, a course of action that sorely distressed Morihei. He was acutely aware of the contradiction between his contention that *budō* was a way of love that fostered peace and preserved life and the massive death and destruction inherent in waging war. In *Budō Renshū*, an instruction manual Morihei produced in 1933, he wrote: "True *budō* is the one that defeats an enemy without sacrificing a single man," and Morihei told his students, "Even in war, the taking of human life is to be avoided as much as possible. Give peace a chance each and every time you are faced with a confrontation." Morihei worked behind the scenes to try to head off a war between the United States and Japan, a subversive act in the eyes of the military police that could have gotten him arrested for treason if found out.

Morihei's message largely fell on deaf ears (after Japan's crushing defeat, one military man confessed, "If we had really understood what Morihei was teaching us we would never have gotten into such a stupid war"), and in

Morihei in 1936, age fifty-three, when he was teaching Aiki-Budō to Japan's military and political elite.

Morihei demonstrating an Aiki-Budō technique in the Noma Dōjō in 1936.

Morihei performing misogi-no-jō, *a rite of purification, before an Aikidō demonstration.*

1942, pleading serious illness, Morihei resigned all his official positions and retired to his farm in Iwama in Ibaraki Prefecture. It was in this year, one of the darkest in human history, that Morihei vowed to dedicate himself to what he now called Aikidō, "the Way of Harmony and the Art of Peace."

In 1945 the war came to its disastrous conclusion for Japan, and the country was in ruins. Morihei later confided to his disciples:

> When the war ended I became deathly ill and fell into a kind of trance. A celestial maiden appeared before me wrapped in flames. I moved toward that brilliant light, but then a Buddhist monk emerged out of the mist and told me, "It is not time for you to go yet. You have a great mission to accomplish on earth."

Morihei recovered his health and announced the new goal for the warriors of Aikidō: "We will train to prevent war, to protect the environment, and to serve society."

Compared with the drama and excitement of his first sixty years, Morihei's final three decades were characterized by a sense of peace and deep spirituality. After the war, Morihei filled his days with Aikidō training, prayer and meditation, the study of religious texts, farmwork, and calligraphy. He once said:

> By secluding myself in Iwama and reducing my involvement in worldly affairs, I have been able to attain a deeper sense of our oneness with nature. I rise each morning at four, purify myself, and then step outdoors to greet the rising sun. I link myself to the cosmos through Aiki and commune with all things—I feel as if I am transformed into the universe itself, breathing in all phenomena. Standing before the altar of heaven and earth, I am in perfect harmony with the Divine. Then I bow in the four directions and pray for universal peace.

Morihei spent much of his time in prayer, meditation, and practice, but he did also manage to travel extensively on instruction tours (including one to Hawaii in 1961). Right to the end of his life, Morihei was constantly refining his technique and expanding his art—"This old man still has to train and train," he told his students.

In 1961, at the age of seventy-eight, Morihei traveled to Hawaii to teach Aikidō and "establish a silver bridge between East and West." This portrait was taken in the Honolulu Headquarters Dōjō.

In 1968 Morihei was hospitalized and his condition was diagnosed as liver cancer. His health gradually declined, but he continued to train and teach as much as he could. Morihei needed assistance to get to and from the dōjō, but once the master stepped onto the mat, he was still the invincible warrior. One disciple related this story:

> Two of us were attending the master on his sickbed one day when he told us, "I'm going to the dōjō." He was not supposed to move around because of his frail condition, but there was no way for us to prevent

Morihei teaching a class in the new Hombu Dōjō a few months before his death. While Morihei's body was weakened by cancer, his ki power remained amazingly powerful—he could down an opponent without touching him, as shown here. Morihei continued training until the very end; he told his students: "Aikidō is a life-long path—continually refine and improve yourselves."

him from doing what he wanted to do. We had to hold him as he walked to the dōjō and carry him up the stairs, but as soon as he stepped onto the mat he straightened up and shrugged his shoulders, a simple movement that somehow sent the two of us flying. The master then placed himself before the dōjō shrine and chanted for a few minutes. After that he trained with us briefly, bowed once more to the shrine, and then sank back gently into our arms as we carried him out of the dōjō and back to his sickroom.

Morihei died peacefully on April 26, 1969. His final words of instruction to his students were: "Aikidō is for the entire world. It is not for selfish or destructive purposes. Train unceasingly for the welfare of all."

About the Photographs
By Walther v. Krenner

I STARTED MY AIKIDŌ TRAINING in 1960 with Isao Takahashi Sensei and Kōichi Tōhei Sensei. Aikidō was brand new in the mainland United States, and Takahashi Sensei had just arrived from Hawaii to introduce Aikidō to the people here. In May of 1963 I received instruction from Kisshōmaru Ueshiba Sensei, Morihei's son and the current head of the Aiki-Kai Federation. I continued my training with Takahashi Sensei until 1967, when he convinced me that I should go to Japan while Morihei (who was known to all as Ō-Sensei, "the Great Master") was still alive.

The same year, armed with a letter of introduction to Ō-Sensei from Takahashi Sensei, I arrived at the old Hombu Dōjō in the Wakamatsu-chō area of Shinjuku. I remember vividly Kisshōmaru Sensei greeting me at the door. Usually he was very formal and reserved, but on this occasion he seemed very friendly and jovial, and I was surprised that he remembered me.

I moved into a dormitory-like building directly across the alley from the dōjō, where several other *gaijin* (foreign) students were housed. I don't remember their names—after all, it was thirty years ago—but I apologize to them for forgetting. I often wonder how many of them are still in Aikidō or even still alive.

Training began the next morning, and on this particular day Ō-Sensei was in the dōjō—he did not teach on a day-to-day basis—and one of the senior students brought me before him; I was introduced and gave him my letter from Takahashi Sensei. A half-hour interview and speech ensued, with Ō-Sensei doing all the talking. I will never forget his voice—it had an ethereal, transcendental quality that engulfed me. I was to start my training on that very day. We trained every day except Saturday and Sunday but had a small class with Morihiro Saitō Sensei on Sunday morning.

After a while I got to know everybody and became friends with some of the other disciples. Sometimes we did things together on days we did not train—I enjoyed hunting for antiques in the countryside—and sometimes we just went sight-seeing. Things were different in those days: there seemed to be a camaraderie and friendship among us, not the politics and rivalries one feels today. None of us had a lot of money, but Japan's economy was much different then—I could manage on five dollars a day.

The photographs in this book are from that time, a time when Ō-Sensei was at the peak of his spiritual quest. Here is a brief account of how the photos came to be: One of the other foreign disciples suggested that we take photographs of Ō-Sensei, since at that time there was only one short volume on Aikidō in English, Kōichi Tōhei's *Aikidō and the Arts of Self-Defense*. He asked me if I was interested in such an undertaking, and I was, so I put some money into the "Ō-Sensei photo project." Ō-Sensei agreed to let us photograph him, but he probably had no idea that the project would turn into a two-hundred-shot photo session that followed him into his home and private living quarters. However, I sensed that he liked to be photographed and did not mind at all.

We used monochrome film because, as I recall, the Japanese color film available in those days did not hold its color well, and monochrome was also cheaper. In retrospect I am glad that things worked out that way, for it gives the photos a classic feel and appearance befitting the subject.

The photos were shot candidly without any posing or script. They portray Ō-Sensei on and off the mat and thus form a natural narrative of a day in the master's life. I believe these photos to be unique, and they clearly stand apart from the other photos taken of Morihei. We gave individual prints to a number of people, including Kisshōmaru, but we did not keep good track of them. Now and then I see one published somewhere and it brings back fond memories of my time at Hombu.

For thirty years I have kept the original complete set together, hoping to share them with others as a total concept in a book such as this. These are the original photos developed in Shinjuku in 1967. As I said, Ō-Sensei no longer taught on a regular schedule at that time, and his illness must have given him great pain; but when he did teach us it was awesome. Sometimes he came in and talked for a long while and left; this was good training in *seiza*. I had the impression that he especially liked to teach foreign students because he always said that Aikidō must build a bridge to the West.

Ō-Sensei would always take the time if someone had a special request. I remember that when I left the Hombu Dōjō to return to Hawaii I wanted to have him autograph a photo for me. The senior students brushed me off, saying that Ō-Sensei was too busy to be bothered, but I persisted, and when Ō-Sensei found out what the senior students had told me, he scolded them for being impolite. Not only did he sign a photo for me, he also presented me with a large calligraphy scroll as a going-away present. That is the last memory I have of him—word reached me that he died a short time later.

Morihei Ueshiba was a *budō* genius and remarkable human being.

Photograph, with calligraphy of "Takemusu Aiki" together with the signature "Ueshiba Morihei," presented to Walther v. Krenner when he left Japan in 1968.

(Left) *Morihei prepares to be filmed in a documentary;* (right) *Morihei and one of his female students performing in the film. Late in life, Morihei liked to be photographed and filmed; no doubt he believed that such a visual testimony would provide future generations the opportunity to see Aikidō in action and would enable them to perceive the truth, goodness, and beauty of the Art of Peace in concrete form.*

Morihei, teaching in the old Hombu Dōjō and telling his students, "Hold the universe in the palm of your hand."

Many of the postures of Morihei captured on film bear a striking resemblance to the postures displayed in classical iconography. Here, Morihei's stance reflects the abhaya-mudrā, the "fear not" gesture of Buddha. Once, Buddha's jealous cousin Devadatta got a royal elephant drunk and goaded the enraged animal to attack Buddha. Just as the elephant was about to crush him, Buddha assumed the "fear not" stance and immediately pacified the beast. Similarly, a verse in the Old Testament refers to this stance: "Thou hast a mighty arm: strong is thy hand, and high is thy right hand" (Psalms 89:13).

PART ONE

ON THE MAT WITH THE MASTER

Shugyō, "diligent training."

"In my Aikidō, there are no opponents, no enemies. I do not want to overwhelm everyone with brute strength, nor do I want to smash every challenger to the ground. In true *budō* there are no opponents, no enemies. In true *budō* we seek to be one with all things, to return to the heart of creation. The purpose of Aikidō training is not to make you simply stronger or tougher than others; it is to make you a warrior for world peace. This is our mission in Aikidō."

(*Above and right*) Morihei bows before the dōjō shrine at the beginning of training. He will make a similar bow at the end of practice—"*Budō* begins and ends with respect."

Morihei always performed some kind of *misogi*, a rite of purification, after bowing in. Usually he did such *misogi* with a wooden sword or staff, but on occasion he used a fan, as photographed here. Morihei said that *misogi* helped him to bridge the gap between the physical and the spiritual realms.

(*Above and following*) A vibrant Morihei warms up with his students. Many of the warm-up movements have practical application as techniques—the movement shown here can be applied in *sokumen-irimi-nage* (pages 30–31).

Morihei taking a deep breath, filling himself with vital energy. Morihei once brushed this calligraphy for a disciple: "Swallow the universe with one gulp!"

In order to practice Aikidō properly, one must "settle down and return to the source." Morihei's meditation here is brief but profound as he calms his spirit and visualizes his mind to be "like the vast sky, the highest peak, and the deepest ocean."

(*Above*) Morihei performs the *tori-fune* (*kogi-fune*) movement, an exercise adopted into Aikidō from traditional Shinto. Morihei stated, "Absorb venerable traditions into Aikidō by clothing them with fresh garments, and build on classic styles to create better forms." (*Right*) Morihei demonstrates the same movement animated by one of his wonderful smiles. Several of his students have stated that they took up the practice of Aikidō after seeing Morihei smile: "Anyone who can smile like that must have really gotten something special from his training."

(*Above and right*) Following the warm-up, Morihei would usually expound on the spiritual significance of Aikidō. These challenging lectures could last for twenty minutes or more.

Morihei holds his *hara*, the body's physical and spiritual center, and talks about the *kototama* (sacred sound) SU, the one-syllable mantra that "fires the blood and creates light, heat, and energy."

The physical techniques of Aikidō typically begin from the *katate-dori*, "one hand held," posture demonstrated here by Morihei.

Morihei executing *shihō-nage*, "four-directions throw," in basic form.

(*Above and following*) Morihei demonstrates the initial entry, the step around, the step in to defuse the attack, and then the final follow-through of the dynamic technique *irimi-nage*, "entering throw." *Irimi-nage* was perhaps Morihei's favorite technique.

(Above and right) Two versions of *sokumen irimi-nage,* "side-entry throw."

(*Above and right*) The *kaiten*, "open and turn," movement lies at the heart of Aikidō techniques.

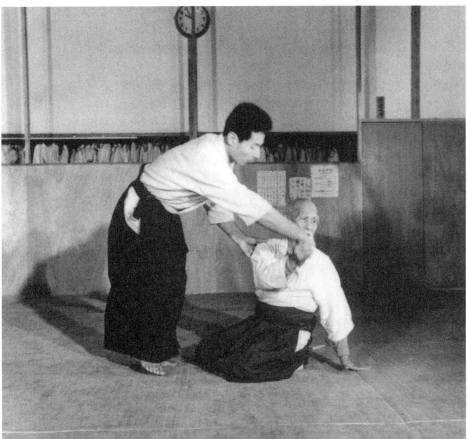

(*Above*) Morihei opens to the side, (*following pages*) and then turns in the same direction as his partner, completely neutralizing the attack.

Morihei explaining the mysteries of *ki*, the vital life force: "Spread your fingertips and let your *ki* flow freely."

Morihei explains, in the following sequence, how to control your partner's *ki* by holding his sleeve.

A suddenly serious and intent Morihei demonstrating the proper method of holding one's partner. Morihei's grip was like a vise and he could immobilize his partner's entire body by simply grasping the person's wrist.

Morihei would frequently demonstrate the close relationship between body movements and sword movements.

Here and on the following pages, Morihei illustrates his explanations of Aikidō with various gestures; the gesture above is reminiscent of the "saint in open prayer" posture often seen in Orthodox Christian iconography.

When Morihei's students would ask him to repeat the movements of a technique he would tell them, "Each and every technique is unique. If I do it again it will naturally be a little different. Rather than trying to grasp how I move, pay attention to what this old fellow is saying to you—that will enable you to execute the techniques properly." When his students protested that his talks were impossibly difficult to comprehend, he replied, "What I tell you may now seem to be way over your heads but eventually there will be a time when it all becomes clear."

Morihei's gestures were naturally much like the movements used in the execution of Aikidō techniques. Here he is both explaining and demonstrating a good Aikidō stance.

In several photographs, Morihei can be seen "pointing out the Way" to his students.

Once a student asked Morihei, "What aspect should we focus on during training?" He answered, "All aspects. In the old martial arts one was taught to 'Perceive each and every inch of an opponent' but in Aikidō we also want to perceive each and every inch of ourselves and our environment."

Morihei demonstrates the hand movements of *kokyū-hō*, "breath power." This kind of vital power emanates from one's center and is superior to mere muscular strength.

(*Above and right*) Morihei executes a graceful *kokyū-nage*, "breath-power throw," his movement as natural and powerful as a large breaking wave.

A jovial Morihei employs *kokyū* power to stymie his partner.

(*Above and following*) More examples of *kokyū-nage*.

Morihei's student had been pressing with all his might against Morihei's forehead to no avail; with a slight turn of his head, Morihei directs him down toward the mat with a *kokyū* throw.

After a *kokyū* throw, Morihei assumes *shizen-tai*, "totally natural posture." In yoga this posture is known as *kayotsarga*, "setting free."

An impish Morihei completely controls his partner with *ikkyō-osae*, "pin number one."

(*Above and following*) Morihei demonstrating variations of *nikyō-osae*, "pin number two."

(*Above and right*) *Sankyō-osae*, "pin number three," employed against a chest hold and combined with a *shihō-nage* variation.

(*Above and right*) Two more examples of pin number three.

On the Mat with the Master · 67

(*Above and right*) A delighted Morihei with his students during practice. Morihei told them, "Aikidō must always be practiced in a joyful and vibrant manner."

(*Above and following*) The concluding exercise in Aikidō is often *suwari kokyū-hō*, "seated breath-power training." Morihei explains the technique in various ways from a variety of

angles. Morihei placed great emphasis on seated techniques and told his disciples to focus on those techniques when Morihei himself could not lead the training.

72 · Training with the Master

The training concludes with the partners giving each other a good back stretch.

Here is how Morihei described the virtues of Aikidō training:

> In order to practice Aikidō properly, you must not forget that all things originate from One Source; envelop yourself with love, and embrace sincerity. A technique that is based only on physical force is weak; a technique based on spiritual power is strong.
>
> The practice of Aikidō is an act of faith, a belief in the power of nonviolence. It is not a type of rigid discipline or empty asceticism. It is a path that follows the principles of nature, principles that must apply to daily living.
>
> In good Aikidō training, we generate light (wisdom) and heat (compassion). Those two elements activate heaven and earth. Train hard and you will experience the light and warmth of Aikidō. Train more, and learn the principles of nature. Aikidō should be practiced from the time you rise to greet the morning sun to the time you retire at night.
>
> Aikidō is good for the health. It helps you manifest your inner and outer beauty. The practice of Aikidō fosters good manners and proper deportment. Aikidō teaches you how to respect others, and how not to behave in a rude manner. It is not easy to live up to the ideals of Aikidō but we must do so at all costs—otherwise our training is in vain.

PART TWO

THE MASTER TEACHES THE WAY

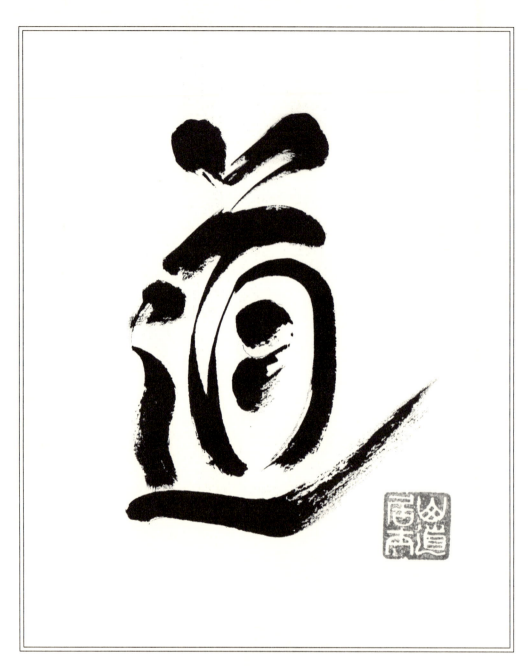

Dō, "the Way."

"The universe is our greatest teacher, our greatest friend. Look at the way a stream wends its way through a mountain valley, smoothly transforming itself as it flows over and around the rocks. The world's wisdom is contained in books, and by studying them, countless new techniques can be created. Study and practice, and then reflect on your progress. Aikidō is the art of learning deeply, the art of knowing oneself."

(*Above and right*) Morihei lecturing on the true nature of Shinto: "Do not think that the Divine exists above us in heaven. The Divine is right here, within and around us. The purpose of Aikidō is to remind us that we are always in a state of grace."

The Master Teaches the Way · 83

(*Above and following*) A relaxed Morihei assumes the "posture of royal ease" (*rajalilasana*) and converses with his students.

(Above and following) Some shots of Morihei standing and talking.

The following sequence shows Morihei receiving visitors in the drawing room. He is shown entering the room, chatting with his guests, receiving a gift, and taking refreshment.

The Master Teaches the Way

Athletes and actors frequently visited Morihei seeking his counsel on how to improve their performances. Sadaharu Oh, the Japanese Babe Ruth, picked up a number of hints from Morihei including this sage advice regarding timing: "Do not try to guess how fast or slow a pitch will be. Just let the ball arrive at its own pace, and be there to greet it." The famous actor Shintarō Katsu asked Morihei how he should portray a blind swordsman in a film: "Create your own universe and bring everything into your own sphere" was the reply.

98 · *Training with the Master*

100 ▸ *Training with the Master*

The Master Teaches the Way

Another sequence taken in Morihei's drawing room.

A final sequence of close-ups of the master Morihei, who never tired of telling his students, "Aikidō is love."

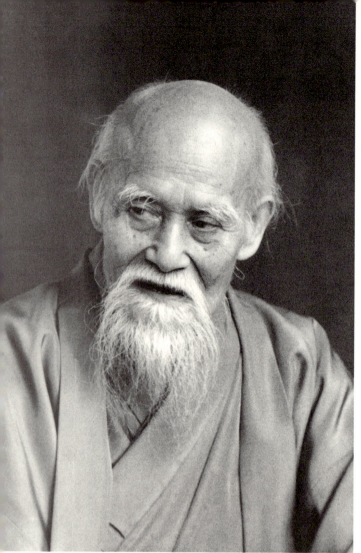

PART THREE

THE ART OF PEACE

Heiwa, "Peace."

The following is a compilation of quotations gathered from Morihei's collected talks, reported sayings, poems, and calligraphy.

THE ART OF PEACE begins with you. Work on yourself and your appointed task in the Art of Peace. Everyone has a spirit that can be refined, a body that can be trained in some manner, a suitable path to follow. You are here for no other purpose than to realize your inner divinity and manifest your innate enlightenment. Foster peace in your own life and then apply the Art to all that you encounter.

, , ,

ONE DOES NOT NEED buildings, money, power, or status to practice the Art of Peace. Heaven is right where you are standing, and that is the place to train.

, , ,

ALL THINGS, material and spiritual, originate from one source and are related as if they were one family. The past, present, and future are all contained in the life force. The universe emerged and developed from one source, and we evolved through the optimal process of unification and harmonization.

The Art of Peace is medicine for a sick world. There is evil and disorder in the world because people have forgotten that all things emanate from one source. Return to that source and leave behind all self-centered thoughts, petty desires, and anger. Those who are possessed by nothing possess everything.

, , ,

If you have not
Linked yourself
To true emptiness,
You will never understand
The Art of Peace.

, , ,

The Art of Peace functions everywhere on earth, in realms ranging from the vastness of space down to the tiniest plants and animals. The life force is all-pervasive and its strength boundless. The Art of Peace allows us to perceive and tap into that tremendous reserve of universal energy.

, , ,

Eight forces sustain creation:
Movement and stillness,
Solidification and fluidity,
Extension and contraction,
Unification and division.

, , ,

Life is growth. If we stop growing, technically and spiritually, we are as good as dead. The Art of Peace is a celebration of the bonding of heaven, earth, and humankind. It is all that is true, good, and beautiful.

Now and again, it is necessary to seclude yourself among deep mountains and hidden valleys to restore your link to the source of life. Breathe in and let yourself soar to the ends of the universe; breathe out and bring the cosmos back inside. Next, breathe up all the fecundity and vibrancy of the earth. Finally, blend the breath of heaven and the breath of earth with that of your own, becoming the Breath of Life itself.

, , ,

All the principles of heaven and earth are living inside you. Life itself is the truth, and this will never change. Everything in heaven and earth breathes. Breath is the thread that ties creation together. When the myriad variations in the universal breath can be sensed, the individual techniques of the Art of Peace are born.

, , ,

Consider the ebb and flow of the tide. When waves come to strike the shore, they crest and fall, creating a sound. Your breath should follow the same pattern, absorbing the entire universe in your belly with each inhalation. Know that we all have access to four treasures: the energy of the sun and moon, the breath of heaven, the breath of earth, and the ebb and flow of the tide.

, , ,

Those who practice the Art of Peace must protect the domain of Mother Nature, the divine reflection of creation, and keep it lovely and fresh. Warriorship gives birth to natural beauty. The subtle techniques of a warrior arise as naturally as the appearance of spring, summer, autumn, and winter. Warriorship is none other than the vitality that sustains all life.

When life is victorious, there is birth; when it is thwarted, there is death. A warrior is always engaged in a life-and-death struggle for Peace.

, , ,

Contemplate the workings of this world, listen to the words of the wise, and take all that is good as your own. With this as your base, open your own door to truth. Do not overlook the truth that is right before you. Study how water flows in a valley stream, smoothly and freely between the rocks. Also learn from holy books and wise people. Everything—even mountains, rivers, plants, and trees—should be your teacher.

, , ,

Create each day anew by clothing yourself with heaven and earth, bathing yourself with wisdom and love, and placing yourself in the heart of Mother Nature.

, , ,

Do not fail
To learn from
The pure voice of an
Ever-flowing mountain stream
Splashing over the rocks.

, , ,

Peace originates with the flow of things—its heart is like the movement of the wind and waves. The Way is like the veins that circulate blood through our bodies, following the natural flow of the life force. If you are separated in the slightest from that divine essence, you are far off the path.

Your heart is full of fertile seeds, waiting to sprout. Just as a lotus flower springs from the mire to bloom splendidly, the interaction of the cosmic breath causes the flower of the spirit to bloom and bear fruit in this world.

, , ,

Study the teachings of the pine tree, the bamboo, and the plum blossom. The pine is evergreen, firmly rooted, and venerable. The bamboo is strong, resilient, unbreakable. The plum blossom is hardy, fragrant, and elegant.

, , ,

Always keep your mind as bright and clear as the vast sky, the great ocean, and the highest peak, empty of all thoughts. Always keep your body filled with light and heat. Fill yourself with the power of wisdom and enlightenment.

, , ,

As soon as you concern yourself with the "good" and "bad" of your fellows, you create an opening in your heart for maliciousness to enter. Testing, competing with, and criticizing others weaken and defeat you.

, , ,

The penetrating brilliance of swords
Wielded by followers of the Way
Strikes at the evil enemy
Lurking deep within
Their own souls and bodies.

THE ART OF PEACE is not easy. It is a fight to the finish, the slaying of evil desires and all falsehood within. On occasion the Voice of Peace resounds like thunder, jolting human beings out of their stupor.

, , ,

CRYSTAL clear,
Sharp and bright,
The sacred sword
Allows no opening
For evil to roost.

, , ,

To practice properly the Art of Peace, you must:
Calm the spirit and return to the source.
Cleanse the body and spirit by removing all malice, selfishness,
  and desire.
Be ever-grateful for the gifts received from the universe, your family,
  Mother Nature, and your fellow human beings.

, , ,

THE ART OF PEACE is based on Four Great Virtues: Bravery, Wisdom, Love, and Friendship, symbolized by Fire, Heaven, Earth, and Water.

, , ,

THE ESSENCE of the Art of Peace is to cleanse yourself of maliciousness, to get in tune with your environment, and to clear your path of all obstacles and barriers.

The only cure for materialism is the cleansing of the six senses (eyes, ears, nose, tongue, body, and mind). If the senses are clogged, one's perception is stifled. The more it is stifled, the more contaminated the senses become. This creates disorder in the world, and that is the greatest evil of all. Polish the heart, free the six senses and let them function without obstruction, and your entire body and soul will glow.

All life is a manifestation of the spirit, the manifestation of love. And the Art of Peace is the purest form of that principle. A warrior is charged with bringing a halt to all contention and strife. Universal love functions in many forms; each manifestation should be allowed free expression. The Art of Peace is true democracy.

Each and every master, regardless of the era or place, heard the call and attained harmony with heaven and earth. There are many paths leading to the top of Mount Fuji, but there is only one summit—love.

Loyalty and devotion lead to bravery. Bravery leads to the spirit of self-sacrifice. The spirit of self-sacrifice creates trust in the power of love.

Economy is the basis of society. When the economy is stable, society develops. The ideal economy combines the spiritual and material, and the best commodities to trade in are sincerity and love.

The Art of Peace does not rely on weapons or brute force to succeed; instead we put ourselves in tune with the universe, maintain peace in our own realms, nurture life, and prevent death and destruction. The true meaning of the term *samurai* is one who serves and adheres to the power of love.

, , ,

Foster and polish
The warrior spirit
While serving in the world;
Illuminate the Path
According to your inner light.

, , ,

The Path of Peace is exceedingly vast, reflecting the grand design of the hidden and manifest worlds. A warrior is a living shrine of the Divine, one who serves that grand purpose.

, , ,

Your mind should be in harmony with the functioning of the universe; your body should be in tune with the movement of the universe; body and mind should be bound as one, unified with the activity of the universe.

, , ,

Even though our path is completely different from the warrior arts of the past, it is not necessary to abandon totally the old ways. Absorb venerable traditions into this new Art by clothing them with fresh garments, and build on the classic styles to create better forms.

Daily training in the Art of Peace allows your inner divinity to shine brighter and brighter. Do not concern yourself with the right and wrong of others. Do not be calculating or act unnaturally. Keep your mind set on the Art of Peace, and do not criticize other teachers or traditions. The Art of Peace never restrains, restricts, or shackles anything. It embraces all and purifies everything.

, , ,

Practice the Art of Peace sincerely, and evil thoughts and deeds will naturally disappear. The only desire that should remain is the thirst for more and more training in the Way.

, , ,

Those who are enlightened never stop forging themselves. The realizations of such masters cannot be expressed well in words or by theories. The most perfect actions echo the patterns found in nature.

, , ,

Day after day
Train your heart out,
Refining your technique:
Use the One to strike the Many!
That is the discipline of a Warrior.

, , ,

The Way of a Warrior
Cannot be encompassed
By words or in letters;
Grasp the essence
And move on toward realization!

THE PURPOSE of training is to tighten up the slack, toughen the body, and polish the spirit.

,   ,   ,

IRON IS FULL of impurities that weaken it; through forging, it becomes steel and is transformed into a razor-sharp sword. Human beings develop in the same fashion.

,   ,   ,

FROM ancient times,
Deep learning and valor
Have been the two pillars of the Path:
Through the virtue of training,
Enlighten both body and soul.

,   ,   ,

INSTRUCTORS CAN IMPART only a fraction of the teaching. It is through your own devoted practice that the mysteries of the Art of Peace are brought to life.

,   ,   ,

THE WAY OF A WARRIOR is based on humanity, love, and sincerity; the heart of martial valor is true bravery, wisdom, love, and friendship. Emphasis on the physical aspects of warriorship is futile, for the power of the body is always limited.

,   ,   ,

A TRUE WARRIOR is always armed with the three things: the radiant sword of pacification; the mirror of bravery, wisdom, and friendship; and the precious jewel of enlightenment.

THE HEART of a human being is no different from the soul of heaven and earth. In your practice always keep in your thoughts the interaction of heaven and earth, water and fire, *yin* and *yang*.

, , ,

THE ART OF PEACE is the principle of nonresistance. Because it is nonresistant, it is victorious from the beginning. Those with evil intentions or contentious thoughts are instantly vanquished. The Art of Peace is invincible because it contends with nothing.

, , ,

THERE ARE no contests in the Art of Peace. A true warrior is invincible because he or she contests with nothing. *Defeat* means to defeat the mind of contention that we harbor within.

, , ,

TO INJURE an opponent is to injure yourself. To control aggression without inflicting injury is the Art of Peace.

, , ,

THE TOTALLY awakened warrior can freely utilize all elements contained in heaven and earth. The true warrior learns how to correctly perceive the activity of the universe and how to transform martial techniques into vehicles of purity, goodness, and beauty. A warrior's mind and body must be permeated with enlightened wisdom and deep calm.

Always practice the Art of Peace in a vibrant and joyful manner.

⁏ ⁏ ⁏

It is necessary to develop a strategy that utilizes all the physical conditions and elements that are directly at hand. The best strategy relies upon an unlimited set of responses.

⁏ ⁏ ⁏

A good stance and posture reflect a proper state of mind.

⁏ ⁏ ⁏

The key to good technique is to keep your hands, feet, and hips straight and centered. If you are centered, you can move freely. The physical center is your belly; if your mind is set there as well, you are assured of victory in any endeavor.

⁏ ⁏ ⁏

Move like a beam of light:
Fly like lightning,
Strike like thunder,
Whirl in circles around
A stable center.

⁏ ⁏ ⁏

Techniques employ four qualities that reflect the nature of our world. Depending on the circumstance, you should be: hard as a diamond, flexible as a willow, smooth-flowing like water, or as empty as space.

I F YOUR OPPONENT strikes with fire, counter with water, becoming completely fluid and free-flowing. Water, by its nature, never collides with or breaks against anything. On the contrary, it swallows up any attack harmlessly.

,   ,   ,

FUNCTIONING harmoniously together, right and left give birth to all techniques. The left hand takes hold of life and death; the right hand controls it. The four limbs of the body are the four pillars of heaven, and manifest the eight directions, *yin* and *yang,* outer and inner.

,   ,   ,

MANIFEST *yang*
In your right hand,
Balance it with
The *yin* of your left,
And guide your partner.

,   ,   ,

THE TECHNIQUES of the Art of Peace are neither fast nor slow, nor are they inside or outside. They transcend time and space.

,   ,   ,

SPRING forth from the Great Earth;
Billow like Great Waves;
Stand like a tree, sit like a rock;
Use the One to strike All.
Learn and forget!

The Art of Peace   ·   127

When an opponent comes forward, move in and greet him; if he wants to pull back, send him on his way.

’ ’ ’

The body should be triangular, the mind circular. The triangle represents the generation of energy and is the most stable physical posture. The circle symbolizes serenity and perfection, the source of unlimited techniques. The square stands for solidity, the basis of applied control.

’ ’ ’

Always try to be in communion with heaven and earth; then the world will appear in its true light. Self-conceit will vanish, and you can blend with any attack.

’ ’ ’

If your heart is large enough to envelop your adversaries, you can see right through them and avoid their attacks. And once you envelop them, you will be able to guide them along a path indicated to you by heaven and earth.

’ ’ ’

Free of weakness,
No-mindedly ignore
The sharp attacks
Of your enemies:
Step in and act!

Do not look upon this world with fear and loathing. Bravely face whatever the gods offer.

, , ,

Each day of human life contains joy and anger, pain and pleasure, darkness and light, growth and decay. Each moment is etched with nature's grand design—do not try to deny or oppose the cosmic order of things.

, , ,

Protectors of this world
And guardians of the Ways
Of gods and buddhas,
The techniques of Peace
Enable us to meet every challenge.

, , ,

Life itself is always a trial. In training, you must test and polish yourself in order to face the great challenges of life. Transcend the realm of life and death, and then you will be able to make your way calmly and safely through any crisis that confronts you.

, , ,

Be grateful even for hardship, setbacks, and bad people. Dealing with such obstacles is an essential part of training in the Art of Peace.

, , ,

Failure is the key to success;
Each mistake teaches us something.

In extreme situations, the entire universe becomes our foe; at such critical times, unity of mind and technique is essential—do not let your heart waver!

At the instant
A warrior
Confronts a foe,
All things
Come into focus.

Even when called out
By a single foe,
Remain on guard,
For you are always surrounded
By a host of enemies.

The Art of Peace is to fulfill that which is lacking.

One should be prepared to receive ninety-nine percent of an enemy's attack and stare death right in the face in order to illumine the Path.

In our techniques we enter completely into, blend totally with, and control firmly an attack. Strength resides where one's *ki* is concentrated and stable; confusion and maliciousness arise when *ki* stagnates.

There are two types of *ki*: ordinary *ki* and true *ki*. Ordinary *ki* is coarse and heavy; true *ki* is light and versatile. In order to perform well, you have to liberate yourself from ordinary *ki* and permeate your organs with true *ki*. That is the basis of powerful technique.

′ ′ ′

In the Art of Peace we never attack. An attack is proof that one is out of control. Never run away from any kind of challenge, but do not try to suppress or control an opponent unnaturally. Let attackers come any way they like and then blend with them. Never chase after opponents. Redirect each attack and get firmly behind it.

′ ′ ′

Seeing me before him,
The enemy attacks,
But by that time
I am already standing
Safely behind him.

′ ′ ′

When attacked, unify the upper, middle, and lower parts of your body. Enter, turn, and blend with your opponent, front and back, right and left.

′ ′ ′

Your spirit is the true shield.

Opponents confront us continually, but actually there is no opponent there. Enter deeply into an attack and neutralize it as you draw that misdirected force into your own sphere.

’ ’ ’

Do not stare into the eyes of your opponent: he may mesmerize you. Do not fix your gaze on his sword: he may intimidate you. Do not focus on your opponent at all: he may absorb your energy. The essence of training is to bring your opponent completely into your sphere. Then you can stand just where you like.

’ ’ ’

Even the most powerful human being has a limited sphere of strength. Draw him outside of that sphere and into your own, and his strength will dissipate.

’ ’ ’

Left and right,
Avoid all
Cuts and parries.
Seize your opponents' minds
And scatter them all!

’ ’ ’

The real Art of Peace is not to sacrifice a single one of your warriors to defeat an enemy. Vanquish your foes by always keeping yourself in a safe and unassailable position; then no one will suffer any losses. The Way of a Warrior, the Art of Politics, is to stop trouble before it starts. It consists in defeating your adversaries spiritually by making them realize the folly of their actions. The Way of a Warrior is to establish harmony.

Master the divine techniques
Of the Art of Peace,
And no enemy
Will dare to
Challenge you.

᠂ ᠂ ᠂

In your training, do not be in a hurry, for it takes a minimum of ten years to master the basics and advance to the first rung. Never think of yourself as an all-knowing, perfected master; you must continue to train daily with your friends and students and progress together in the Art of Peace.

᠂ ᠂ ᠂

Progress comes
To those who
Train and train;
Reliance on secret techniques
Will get you nowhere.

᠂ ᠂ ᠂

Fiddling with this
And that technique
Is of no avail.
Simply act decisively
Without reserve!

IF YOU PERCEIVE the true form of heaven and earth, you will be enlightened to your own true form. If you are enlightened about a certain principle, you can put it into practice. After each practical application, reflect on your efforts. Progress continually like this.

*   *   *

THE ART OF PEACE can be summed up like this: *True victory is self-victory; let that day arrive quickly!* "True victory" means unflinching courage; "self-victory" symbolizes unflagging effort; and "let that day arrive quickly" represents the glorious moment of triumph in the here and now.

*   *   *

CAST OFF limiting thoughts and return to true emptiness. Stand in the midst of the Great Void. This is the secret of the Way of a Warrior.

*   *   *

TO TRULY IMPLEMENT the Art of Peace, you must be able to sport freely in the manifest, hidden, and divine realms.

*   *   *

IF you comprehend
The Art of Peace,
This difficult path,
Just as it is,
Envelops the circle of heaven.

The techniques of the Way of Peace change constantly; every encounter is unique, and the appropriate response should emerge naturally. Today's techniques will be different tomorrow. Do not get caught up with the form and appearance of a challenge. The Art of Peace has no form—it is the study of the spirit.

, , ,

Ultimately, you must forget about technique. The further you progress, the fewer teachings there are. The Great Path is really No Path.

, , ,

The Art of Peace that I practice has room for each of the world's eight million gods, and I cooperate with them all. The God of Peace is very great and enjoins all that is divine and enlightened in every land.

, , ,

The Art of Peace is a form of prayer that generates light and heat. Forget about your little self, detach yourself from objects, and you will radiate light and warmth. Light is wisdom; warmth is compassion.

, , ,

Construction of shrine and temple buildings is not enough. Establish yourself as a living buddha image. We all should be transformed into goddesses of compassion or victorious buddhas.

Rely on Peace
To activate your
Manifold powers;
Pacify your environment
And create a beautiful world.

’ ’ ’

The Divine is not something high above us. It is in heaven, it is in earth, it is inside us.

’ ’ ’

Unite yourself to the cosmos, and the thought of transcendence will disappear. Transcendence belongs to the profane world. When all trace of transcendence vanishes, the true person—the Divine Being—is manifest. Empty yourself and let the Divine function.

’ ’ ’

You cannot see or touch the Divine with your gross senses. The Divine is within you, not somewhere else. Unite yourself to the Divine, and you will be able to perceive gods wherever you are, but do not try to grasp or cling to them.

’ ’ ’

The Divine does not like to be shut up in a building. The Divine likes to be out in the open. It is right here in this very body. Each one of us is a miniature universe, a living shrine.

When you bow deeply to the universe, it bows back; when you call out the name of God, it echoes inside you.

, , ,

The Art of Peace is the religion that is not a religion; it perfects and completes all religions.

, , ,

The Path is exceedingly vast. From ancient times to the present day, even the greatest sages were unable to perceive and comprehend the entire truth; the explanation and teachings of masters and saints express only part of the whole. It is not possible for anyone to speak of such things in their entirety. Just head for the light and heat, learn from the gods, and through the virtue of devoted practice of the Art of Peace, become one with the Divine.

Suggestions for Further Reading

FOR A BASIC INTRODUCTION to the art of Aikidō see my *Shambhala Guide to Aikidō* (Boston: Shambhala Publications, 1996). Details of Morihei's life and times are found in my *Invincible Warrior: An Illustrated Biography of Morihei Ueshiba, Founder of Aikidō* (Boston: Shambhala Publications, 1997). A video biography of Morihei Ueshiba, in six volumes, has been produced by Aiki News. Two books published under Morihei Ueshiba's direction are available in English translation: *Budō Training in Aikidō* (Tokyo: Sugawara Martial Arts Institute, 1997) and *Budō* (Tokyo: Kōdansha International, 1991). For in-depth studies of Morihei's philosophy see *The Essence of Aikidō: Spiritual Teachings of Morihei Ueshiba*, compiled by John Stevens (Tokyo: Kōdansha International, 1993), and my *Secrets of Aikidō* (Boston: Shambhala Publications, 1995). Good general books on Aikidō include *Aikidō* by Kisshōmaru Ueshiba (Tokyo: Hōzansha, 1985); two books by Mitsugi Saotome, *The Principles of Aikidō* and *Aikidō and the Harmony of Nature*, both published in 1993 by Shambhala Publications; *Ultimate Aikidō* by Yoshimitsu Yamada (New York: Citadel Press, 1994); *Aikidō and the Dynamic Sphere* by A. Westbrook and O. Ratti (Rutland, Vt.: Tuttle, 1970); and *Aikidō, the Way of Harmony* by John Stevens under the direction of Rinjirō Shirata (Boston: Shambhala Publications, 1984).